GOBLIN SHARKS

MARYSA STORM

Bolt Jr. is published by Black Rabbit Books
P.O. Box 227, Mankato, Minnesota, 56002
www.blackrabbitbooks.com

Michael Sellner, designer and photo researcher

Library of Congress Cataloging-in-Publication Data
Names: Storm, Marysa, author.
Title: Goblin sharks / by Marysa Storm.
Description: Mankato: Black Rabbit Books, [2024] |
Series: Bolt Jr. World of sharks | Includes bibliographical references and index. |
Audience: Ages 6–8 | Audience: Grades K–1 |
Summary: "Goblin sharks are mysterious animals and fearsome eating machines. Let beginning readers dive in and learn all about these creatures through closely leveled text, vibrant imagery, and captivating infographics"—
Provided by publisher.
Identifiers: LCCN 2021030408 (print) | LCCN 2021030409 (ebook) | ISBN 9781623107895 (hardcover) |
ISBN 9781623107956 (ebook)
Subjects: LCSH: Goblin shark—Juvenile literature.
Classification: LCC QL638.95.M58 S76 2024 (print) |
LCC QL638.95.M58 (ebook) | DDC 597.3—dc23
LC record available at https://lccn.loc.gov/2021030408
LC ebook record available at https://lccn.loc.gov/2021030409

Image Credits
Alamy: Animal Stock 6–7, Kelvin Aitken/VWPics 20–21; Blue Planet Archive: David Shen 18, Makoto Hirose/e-Photo 10, 13, Marko Steffensen 12, Masa Ushioda 8–9; Diane Bray/Museum Victoria cover; HOKKAIDO UNIVERSITY: Okinawa Churashima Foundation 4; Museums Victoria: Dianne J. Bray 1, 23; Shutterstock: 3DMI 7, designer_an 3, 24, Gemma Ellen 14–15, Vladimir Arndt 17, Yongcharoen_kittiyaporn 5, 23; Tumblir: irlShitposter 17; Twitter: Science Channel 11

Contents

CHAPTER 1

Swimming Along

Deep in the ocean, a strange fish swims. It has soft skin. Its **snout** is long. Sharp teeth fill its mouth. This strange animal is a goblin shark.

snout: the long front part of an animal's head that includes the nose, mouth, and jaws.

COMPARING LENGTHS
goblin shark
up to about 12.5 feet
(4 meters)

Jaws

These sharks have interesting features. Their jaws are extra strange. The jaws shoot forward when the sharks attack.

PARTS OF A

Goblin Shark

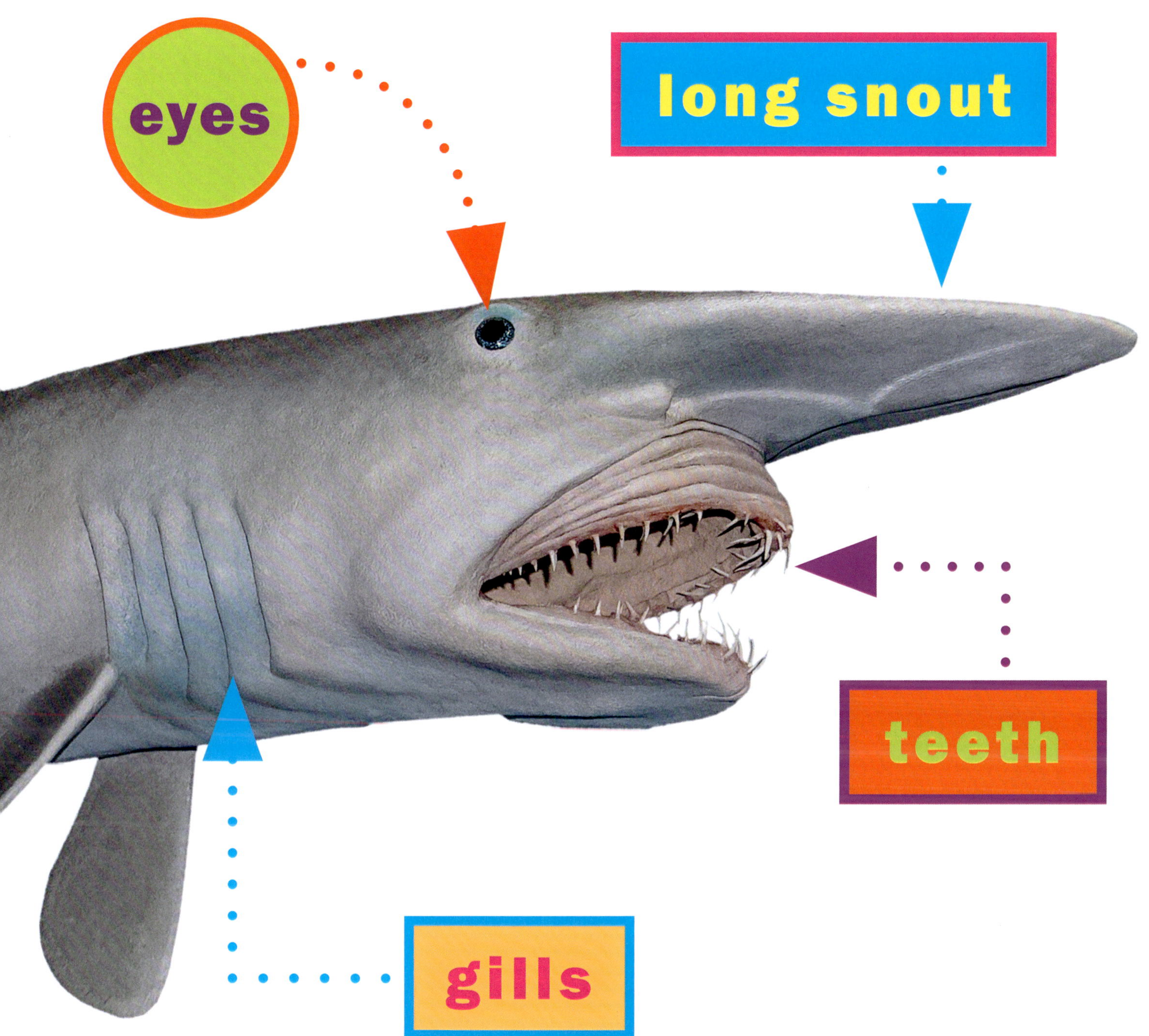
eyes
long snout
teeth
gills

CHAPTER 2

Food and Homes

Goblin sharks have pointy teeth. They're perfect for catching **prey**. These sharks eat fish. They also eat crabs and squid.

prey: an animal hunted or killed for food

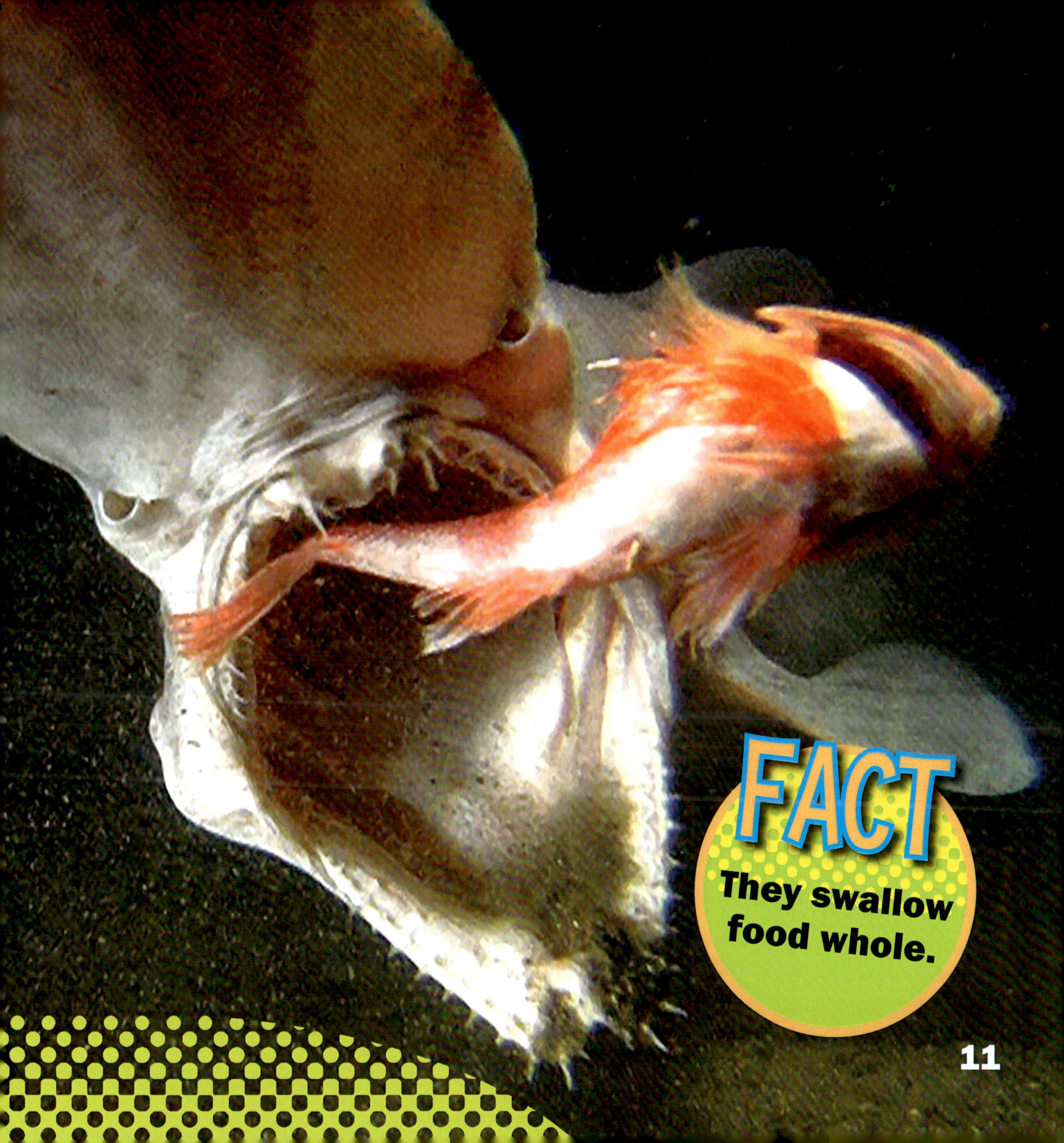
FACT
They swallow food whole.

Where They Swim

These sharks live deep in the ocean. They can swim 4,265 feet (1,300 m) down. It is dark down there. The sharks' snouts help them find food. This depth makes them hard to **study**.

study: to learn about a subject

Where Goblin Sharks Have Been Found

North America

South America

KEY

= where goblin sharks have been found

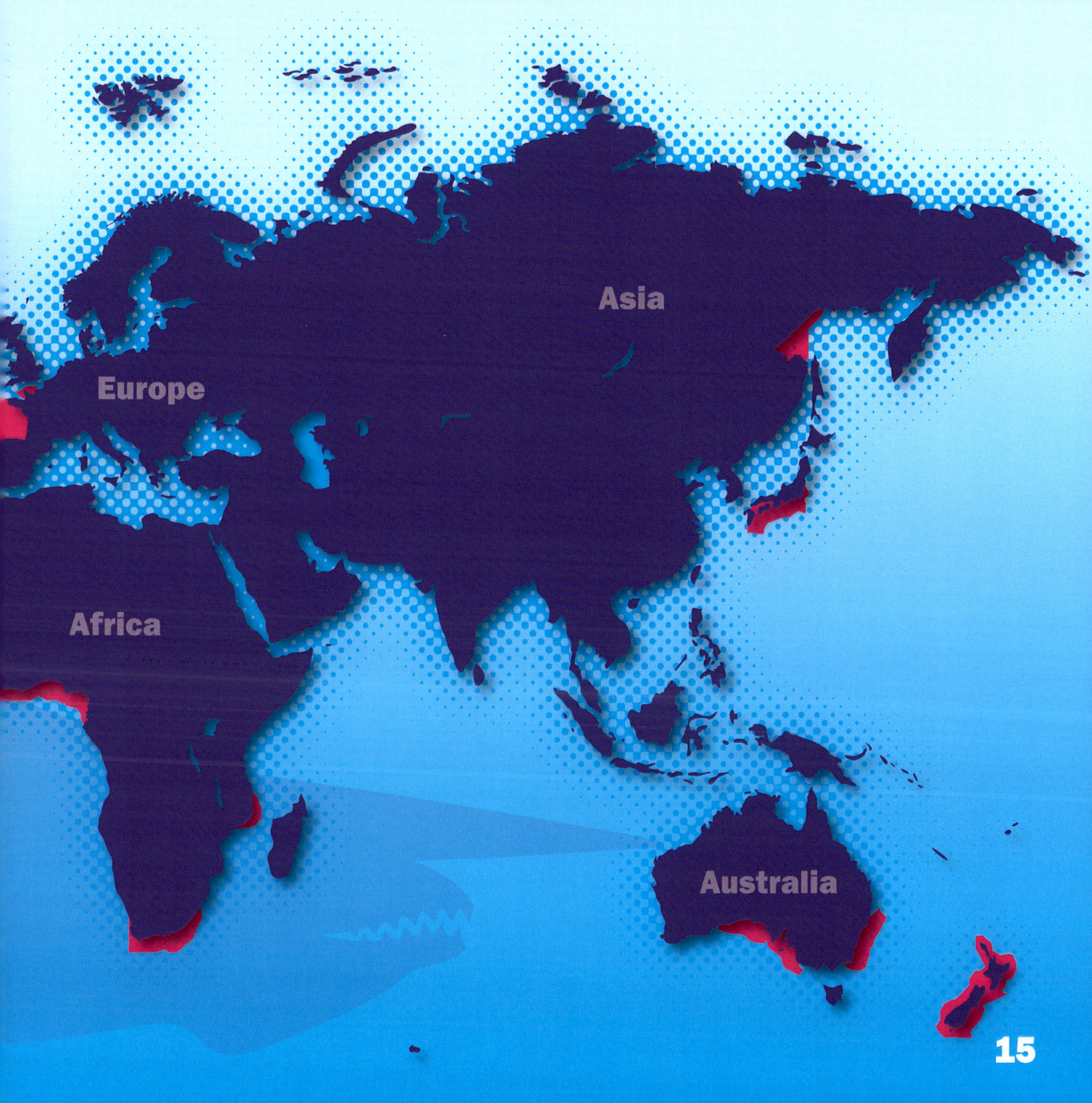
Asia
Europe
Africa
Australia

CHAPTER 3

Family Life

People don't know a lot about how these sharks act. Most sharks live alone, though. Goblins probably do too. They meet to have babies.

Learning More

People don’t know how many babies these sharks have. They’re not sure how long the adults live. Most goblins studied by scientists are dead. People hope to study live goblin sharks. Underwater robots will help.

length of the smallest goblin shark ever found

3.5 feet
(1 m)

Bonus Facts

Goblins have 10 gills.

Their jaws move forward at **10 feet** (3 m) per second.

People discovered the sharks in 1898.

Humans are sharks' biggest **threat**.

threat: something that can hurt someone or something

READ MORE/WEBSITES

Hansen, Grace. *Goblin Sharks.* Spooky Animals. Minneapolis: Abdo Kids Jumbo, 2021.

Lopetz, Nicola. *Sharks!: Big Teeth, Fierce Hunters.* Built to Survive. New York: Crabtree Publishing, 2022.

Siemens, Jared. *Sharks.* Creatures of the Sea. New York: AV2, 2022.

Goblin Shark
kids.nationalgeographic.com/animals/fish/facts/goblin-shark

Goblin Shark
oceana.org/marine-life/goblin-shark

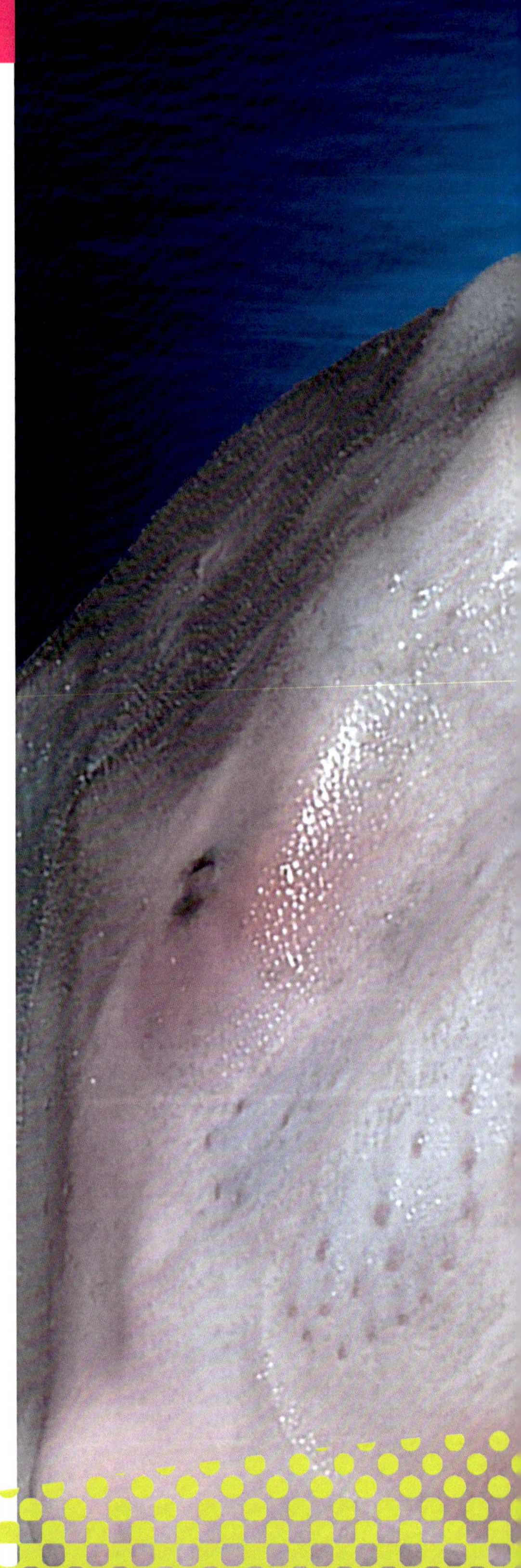